Geography

FOR COMMON ENTRANCE

13+

Exam Practice Answers

Belinda Froud-Yannic

About the author

Belinda Froud-Yannic has been Head of Geography at Thomas's School, Clapham since 2001. She formerly taught at Broomfield School, Southgate. Most of her spare time is taken up with looking after her two children but she also tries to squeeze in some pottery, hill walking and skiing. She believes geography is a subject that everyone can enjoy due to its diversity of themes and that lessons should be fun and full of fieldwork!

The publishers would like to thank the following for permission to reproduce copyright material:
Photo credits p19 © geogphotos / Alamy

Every effort has been made to trace all copyright holders, but if any have been inadvertently overlooked the publishers will be pleased to make the necessary arrangements at the first opportunity.

Although every effort has been made to ensure that website addresses are correct at time of going to press, Galore Park cannot be held responsible for the content of any website mentioned in this book. It is sometimes possible to find a relocated web page by typing in the address of the home page for a website in the URL window of your browser.

Hachette UK's policy is to use papers that are natural, renewable and recyclable products and made from wood grown in sustainable forests. The logging and manufacturing processes are expected to conform to the environmental regulations of the country of origin.

Orders: please contact Bookpoint Ltd, 130 Milton Park, Abingdon, Oxon OX14 4SB. Telephone: +44 (0)1235 827827. Lines are open 9.00a.m.–5.00p.m., Monday to Saturday, with a 24-hour message answering service. Visit our website at www.galorepark.co.uk for details of other revision guides for Common Entrance, examination papers and Galore Park publications.

Published by Galore Park Publishing Ltd
An Hachette UK company
Carmelite House, 50 Victoria Embankment, London, EC4Y 0DZ
www.galorepark.co.uk

Impression number 10 9 8 7 6 5
2018 2017

Typeset in India
Printed in the UK
Illustrations by Integra Software Services Pvt. Ltd., Pondicherry, India and Aptara, Inc.

A catalogue record for this title is available from the British Library.

ISBN: 978 1 471827 32 7

Contents

Introduction

This book provides a full set of answers to the questions in *Geography for Common Entrance 13+ Exam Practice Questions* by Belinda Froud-Yannic, ISBN 9781471827310 (2014).

Teachers and parents are advised to use their own discretion in judging whether a pupil has answered a question completely and accurately enough to be awarded full marks. A pupil may arrive at a unique answer not accounted for in this book, and these answers should not necessarily be discounted.

The Geography Common Entrance exam

The Geography Common Entrance exam is one hour long. The paper is split into three sections: location knowledge, Ordnance Survey and thematic studies. You should answer all the questions in all three sections.

The location knowledge and Ordnance Survey sections are each worth 10–15 marks and the thematic studies section is worth 50–60 marks. Another 20 marks come from your fieldwork investigation. This makes a total of 100 marks or 100 percent. Your Common Entrance paper will be marked by the senior school that you hope to attend. The school will work out your final percentage and turn this into a grade (A, B, C, etc.). The percentage required to obtain a particular grade differs between schools.

Location knowledge

- It is best to start with this section of the exam as it can be completed quickly and easily if you have learnt your locations.
- You should spend about **eight minutes** on this section.
- Make sure that you read the questions carefully. If, for instance, you are asked for the name of a country do *not* write the name of a city!
- If you are asked to mark something on a map, such as a line of latitude or a mountainous area, do not forget to label it.
- Make sure that you practise marking the locations on the continent and world maps in this book.
- This is the most straightforward section to revise for as it is just a case of learning and practising. Quizzes with your family and friends will also help you to revise.

Ordnance Survey

- Ensure that you have a sharp pencil, a ruler and a scrap of paper or a piece of string.
- It is important that you have a flat surface onto which you can place the OS map. You may need to move some items from your desk onto the floor.
- You should spend **10–12 minutes** on this section.
- Make sure that you read the instructions carefully and double check all your answers. If the question asks for a distance, do *not* give a direction as your answer!
- Ensure that you always add the correct units to any answer. Use kilometres (km) for distance and metres (m) for altitude.

- Give a six-figure reference for any spot (small) features such as a post office or milestone but a four-figure reference for large features such as woodlands or towns.
- Ensure that you look carefully at the word 'from' in a direction question so that you do not 'go' the wrong way.
- If you are asked to describe a route, remember to break the route into sections and give altitudes, directions and distances, and mention any features that you pass.

Thematic studies

- You should spend **35–40 minutes** on this section.
- There will be questions from each of the five themes: earthquakes and volcanoes, weather and climate, rivers and coasts, population and settlement, and transport and industry.
- Some questions may refer back to the OS map; other questions may use resources such as photos, graphs or diagrams. You must study these carefully before answering the question. (Remember that the line on a climate graph is the temperature and the blocks are the rainfall.)
- You will be given marks for including examples and for drawing relevant diagrams, even if the question does not specifically ask you to do this.
- If you are asked a question about a case study, make sure that you make your answer specific by using names of places and actual figures.

General points

- Have a watch on your desk. Work out how much time you need to allocate to each question and try to stick to it.
- Make sure you read and understand the instructions and rules on the front of the exam paper.
- Always read the questions carefully, underlining, circling or highlighting key words or phrases.
- Look at the number of marks available in order to assess how much to write for each answer. If you use bullet points to answer a question that offers a high mark, you must make sure that the bullet points include sufficient detail.
- Do not leave blanks. If you do not know the answer, take an educated guess. Wrong answers do not lose marks.
- Make sure that all your diagrams are clearly annotated (labelled with explanations). There are certain diagrams that it is essential you know how to draw. These are clearly marked throughout this book.
- Whenever possible, include impressive geographical terms from the lists of 'words you need to know'. This creates a good impression and will gain you higher marks.
- If a question is particularly hard move on to the next one. Go back to it if you have time at the end.
- Organise your time so that you have time to check your answers at the end.

Command words

Make sure you completely understand these words and phrases. Cover up the definitions with a sheet of paper in order to test yourself.

annotate	add descriptive explanatory labels
choose	select carefully from a number of alternatives
complete	finish, make whole
define	give an exact description of
describe	write down the nature of the feature
develop	expand upon an idea
explain	write in detail how something has come into being and/or changed
give	show evidence of
identify	find evidence of
list	put a number of examples in sequence
mark and name	show the exact location of and add the name
name	give a precise example of
select	pick out as most suitable or best
shade and name	fill in the area of a feature and add the name
state	express fully and clearly in words
study	look at and/or read carefully
suggest	propose reasons or ideas for something

These words are only used in the scholarship exam:

discuss	present viewpoints from various aspects of a subject
elaborate	similar to expand and illustrate
expand	develop an argument and/or present greater detail on
illustrate	use examples to develop an argument or theme

➜ Tips on taking the exam

Before the exam

- Have all your equipment and pens ready the night before. You will need: ruler, calculator, red, yellow and blue colouring pencils, two normal pencils, an ink pen and spare cartridges.
- Make sure you are at your best by getting a good night's sleep before the exam.
- Have a good breakfast in the morning.
- Take some water into the exam if you are allowed.
- Think positively and keep calm.

1 Earthquakes and volcanoes

1 (a) mantle (1)

(b) a constructive boundary is formed (1)

(c) plates move towards each other (1)

(d) destructive and constructive boundaries (1)

(e) a destructive plate boundary is formed (1)

(f) volcanoes and earthquakes are formed (1)

(g) destructive plate boundary (1)

2 (a) ocean trench (1)

(b) destructive (1)

(c) magma (1)

(d) liquid (1)

(e) seismometer (1)

(f) convection currents (1)

3 (a) occur (1)

(b) lighter (1)

(c) fold mountains (1)

(d) conservative/sliding (1)

(e) granite (1)

4 (a) *One mark for each point and an extra mark if the point is developed, to a maximum of four marks.*

- They are in narrow bands.
- There is a band down the middle of the Atlantic.
- There is a band around the edge of the Pacific Ocean.
- There is a band through the Mediterranean and central Asia. (4)

(b) (i) The Mid Atlantic Ridge (1)

(ii) The Pacific Ring of Fire (1)

(c) *One mark for each point and an extra mark if the point is developed, to a maximum of five marks.*

- Volcanoes occur at destructive and constructive plate boundaries.
- Volcanoes occur at destructive plate boundaries as the oceanic plate sinks under the continental plate and the oceanic plate melts and the magma rises as it is full of gas bubbles.
- Volcanoes occur at constructive plate boundaries as the oceanic plates move apart and the magma is allowed to escape.
- Earthquakes occur at plate boundaries where the plates rub and create friction.
- Plates could lock and then when the pressure is released an earthquake could occur. (5)

5 (a) A tectonic plate is a huge slab of rock which makes up the Earth's crust and which floats on the mantle. (1)

(b) A tsunami is a tidal wave which could be caused by the movement of plates (an earthquake). (1)

(c) A seismic wave is a shock wave produced by an earthquake. (1)

(d) The epicentre is the point on the Earth's surface directly above the focus of the earthquake and is where the earthquake is felt most strongly. (1)

6 (a) *Answer depends on the choice of volcanic eruption.* (2)

(b) *One mark for each point and an extra mark if the point is developed. Extra marks can be awarded for any diagrams used to illustrate the point, up to a maximum of five marks.*

If pupil has chosen an eruption on a destructive plate boundary, the following should be mentioned:

- Oceanic plate meets continental plate.
- Oceanic plate subducts under continental plate.
- Oceanic plate melts.
- Excess magma rises as it is full of gas bubbles.
- A volcano is formed in the fold mountains.

If pupil has chosen an eruption on a constructive plate boundary, the following should be mentioned:

- Oceanic plates move apart.
- Magma can escape from the mantle where the plates move apart. (5)

(c) *One mark for each point and an extra mark if the point is developed, to a maximum of four marks. Answer depends on the choice of volcanic eruption but could cover the following:*

- loss of life
- loss of property
- tourism economy affected

- agriculture economy affected
- communications damaged
- respiratory problems
- services such as schools and hospitals closed
- evacuation. (4)

(d) *One mark for each point and an extra mark if the point is developed, to a maximum of four marks. Answer depends on the choice of volcanic eruption but could cover the following:*

- pyroclastic flow destroying vegetation, buildings, etc.
- agricultural land destroyed
- lava could flow into sea and form new land
- communications destroyed
- ash could kill wildlife on land and in the sea. (4)

7 *One mark for each point and an extra mark if the point is developed, to a maximum of four marks.*

In less economically developed countries:

- safety regulations are not in force for buildings, which can therefore collapse more easily
- the fertile volcanic cone is likely to be densely populated as many people are subsistence farmers
- the emergency services are either non existent or not as well equipped as in more economically developed countries
- the likelihood of prediction is lower
- hospitals are less likely to be able to cope with the influx of patients
- access to clean water for washing and drinking may be a problem which may cause the spread of disease as a secondary effect
- evacuation is difficult as fewer vehicles are available and roads are poor. (4)

8 **(a)** Destructive plate boundary (4)

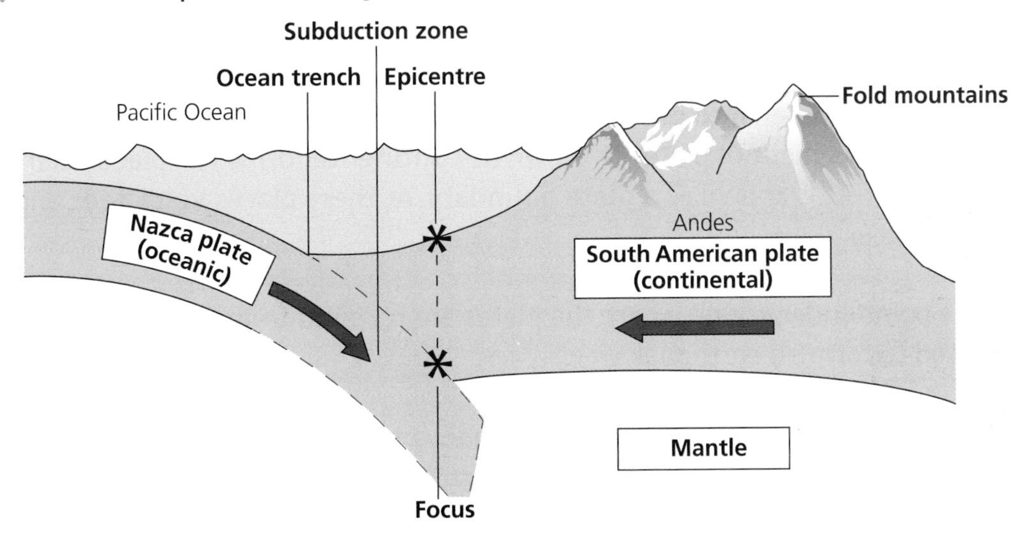

(b) *Arrows marked correctly on diagram to show movement of plates* (2)

9 (a) Due to convection currents (1)

(b) European and Messina plates (1)

(c) Destructive because the plates are moving towards each other (2)

10 (a) At a destructive plate boundary (1)

(b) A destructive plate boundary (1)

(c) A constructive plate boundary (1)

(d) A conservative/sliding plate boundary (1)

(e) A collision plate boundary (1)

(f) Continental drift (1)

11 (a) cause (1)

(b) cause (1)

(c) effect (1)

(d) effect (1)

(e) effect (1)

12

A volcano which will never erupt again	extinct
A volcano which could erupt but has not erupted recently	dormant
A volcano which is erupting or showing signs of activity	active

(3)

13 The Earth is made up of layers. The centre of the Earth is known as the **inner core**. Outside of this layer is the outer core. Around this is the **mantle** which is made up of molten **magma**. This liquid rock is continuously moving as **convection currents**. The outermost layer is called the **crust**. This is broken into **plates** which gradually move due to the magma moving beneath them. This movement of plates is known as **continental drift**. The place where two or more plates meet is known as a **plate boundary**. At these places **volcanoes** and **earthquakes** can occur. (10)

14 Focus: the exact point underground where the plates are released from their lock and seismic waves are created to form an earthquake (1)

Epicentre: the point on the Earth's surface directly above the focus of an earthquake (1)

15 *One mark for each point and an extra mark if the point is developed, to a maximum of four marks.*

People live in areas that are prone to earthquakes and volcanoes:

- to use the fertile soil for agriculture
- to use the geothermal energy
- to make money from tourists who visit the volcano
- to mine for minerals and gem stones. (4)

16 *One mark for each point and an extra mark if the point is developed, to a maximum of four marks. Specific examples must be given in order to score full four marks.*

- Build earthquake-proof buildings.
- Build cities away from plate boundaries.
- Use computers that turn off the gas supply when an earthquake occurs.
- Predict volcanoes using tiltmeters and seismometers.
- Create tsunami warning systems.
- Divert lava flow.
- Create hazard maps to evacuate areas after an eruption.
- Avoid building on or near volcanic cones.
- In Japan some buildings are designed by architects to withstand earthquakes. They have rubber in their foundations and also have wide bases and narrow tops.
- In Montserrat they have a volcano observatory which monitors the volcano using seismometers and tiltmeters to assess when it is going to erupt. (4)

17 *One mark for each point and an extra mark if the point is developed, to a maximum of four marks.*

Lava flow, pyroclastic flow and lahars from volcanic eruptions and earthquakes can cause:

- loss of life
- loss of business
- loss of communications
- damage to property and farmland
- tsunamis (earthquakes). (4)

18 *One mark for each point and an extra mark if the point is developed, to a maximum of three marks.*

- Population density
- Level of development of country
- Time – level of technology for prediction increases over time
- Power/magnitude of volcano/earthquake – relating to type of plate boundary (3)

19 (a) *One mark for each point and an extra mark if the point is developed, to a maximum of four marks. Answer depends on the example chosen, but should include the following words:*

- plate boundary
- friction
- tension
- locking. (4)

(b) *One mark for each point and an extra mark if the point is developed, to a maximum of three marks. Answer depends on the example chosen, but could include the following words:*

- aid
- rescue forces
- evacuation
- better building methods in the future. (3)

(c) *One mark for each point and an extra mark if the point is developed, to a maximum of six marks. Answer depends on the example chosen, but could include:*

- loss of life
- loss of property
- loss of tourism
- businesses losing money
- communications destroyed
- spread of disease
- services destroyed
- looting. (6)

20 *One mark for each point and an extra mark if the point is developed, to a maximum of three marks.*

In more economically developed countries the:

- likelihood of accurate prediction is greater
- likelihood of evacuation is greater
- quality of roads is better
- building materials and the technology put into architecture are likely to be better
- quality of aftercare in hospital for the injured is likely to be better.

In less economically developed countries there could be a large number of subsistence farmers living around a volcanic cone. (3)

2 Weather and climate

1 (a) in mountainous areas (1)

(b) in rainforests every day (1)

(c) when a warm air mass rises over a cold air mass (1)

(d) rain, snow, hail or sleet (1)

(e) when rain hits a roof or tree before landing (1)

(f) the vertical movement of water through soil (1)

(g) cooling turns water vapour into a liquid (1)

(h) the height above sea level (1)

(i) in the east of England (1)

(j) the direction something is facing (1)

(k) a current of air high in the atmosphere (1)

(l) south (1)

2 (a) Site 4 (1)

(b) Site 3 (1)

(c) Any one of:

- The dark surface absorbs heat.
- There is no shade.
- It is at low altitude.
- It is sheltered from the prevailing wind. (1)

(d) Site 1 (1)

(e) Any one of:

- It is at high altitude.
- There is no shelter from the prevailing wind. (1)

(f) Site 1 (1)

(g) Any one of:

- It is at high altitude.
- There is no shelter from the prevailing wind. (1)

3 Relief rainfall (5)

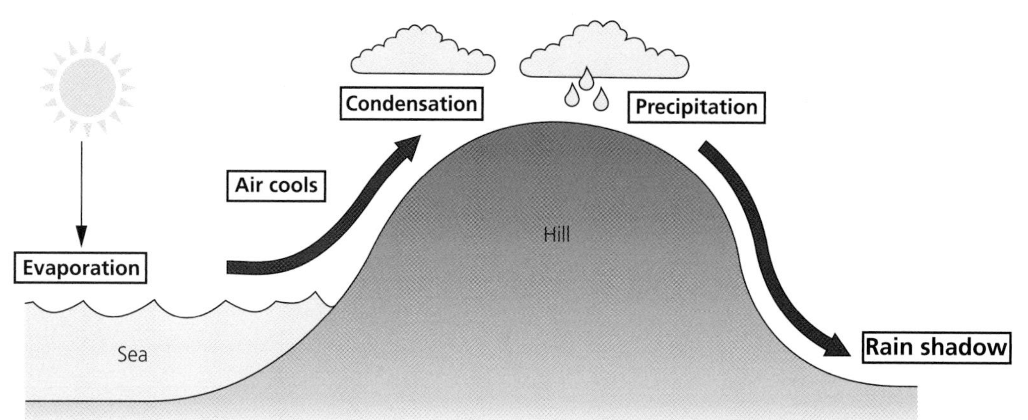

4 *Two explanations only and a maximum of two marks for each. One mark can be given for each point and two marks if the point is developed. One mark can be awarded for names of places in the British Isles used as examples.*

Altitude

- The higher the altitude the lower the temperature.
- The higher the altitude the higher the wind speed.
- The higher the altitude the more likely it is to snow.
- Places at high altitude are more likely to receive relief rainfall.
- Therefore Ben Nevis is colder, wetter and windier than Edinburgh. (2)

Latitude

- The closer to the Equator the higher the temperature.
- The closer to the Equator the more likely the place is to receive convectional rainfall.
- Therefore Bournemouth is warmer and more likely to have convectional rainfall than Edinburgh. (2)

Distance from the sea

- The closer to the sea in summer the higher the temperature.
- The closer to the sea in winter the colder the temperature.
- Places near the sea are likely to be windier due to lack of shelter. (2)

5 Convectional rainfall (5)

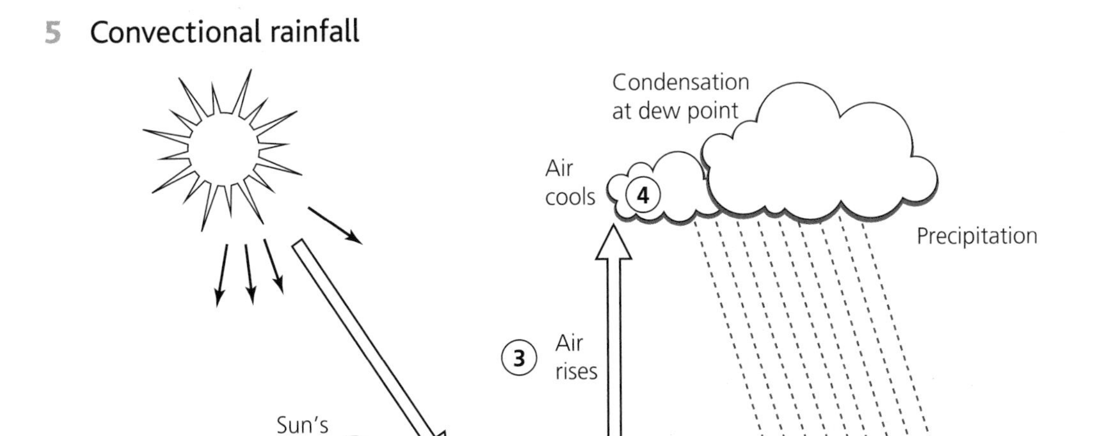

6 **(a)** Humidity is the percentage of water vapour in the air. (1)

(b) Transpiration is the release of water vapour from the leaves of vegetation. (1)

(c) Prevailing wind is the wind from the most usual (predominant) direction. (1)

(d) Surface run-off is the movement of water over the surface of the ground. (1)

7 **(a)** Microclimate is the climate of a small area. (1)

(b) Any three of:

- aspect
- physical features (hills, lakes, etc.)
- buildings
- colour of surface. (3)

(c) warmer (especially at night), lower wind speed and less snow (3)

(d)
- Buildings (bricks) absorb heat during the day and release it slowly at night.
- Tarmac absorbs heat.
- Buildings create shelter which reduces wind speed.
- Cars and central heating systems release heat. (3)

8 **(a)** convectional (1)

(b) west (1)

(c) vertical (1)

(d) warms (1)

(e) warmer (1)

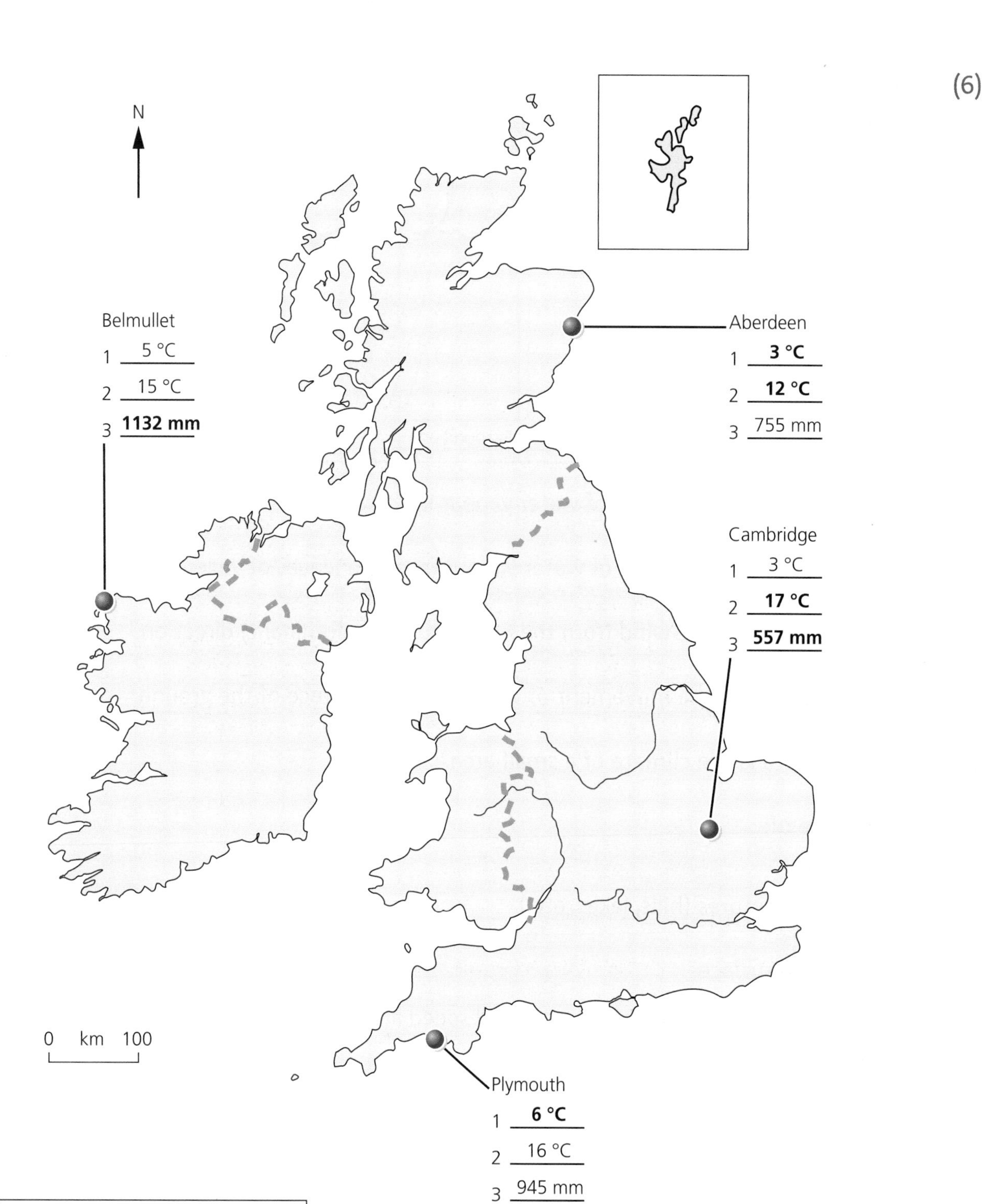

(6)

10 (a) south-west **(1)**

(b) *One mark can be given for each point and two marks if the point is developed. One mark can be awarded for names of places in the British Isles used as examples.*

The prevailing wind:

- blows across Atlantic bringing warm, moist air
- brings moist air which condenses as the air is forced up mountains in west

- brings relief rainfall to the west of the country
- brings warmth to the west of the country as it is from the south
- brings warm air masses which meet the cold air masses from the north and therefore create frontal rainfall. (4)

11 ● 2991 (1)

● It is at lower altitude. (1)

12 Frontal rainfall (5)

13 Any two of:

Weather:

● is day to day

● is the condition of the atmosphere

● includes temperature, precipitation and wind speed

● is short term whereas climate is described over a longer period of time. (1)

Climate:

● is the average weather for a place

● is usually described over one year

● is average conditions whereas weather is actual conditions. (1)

14 (a) July (1)

(b) January (1)

(c) November (1)

(d) 15 °C (1)

(e) *One mark can be given for each point and two marks if the point is developed. One mark can be awarded for names of places in the British Isles used as examples. Up to a maximum of six marks.*

The climate of the north-west of the British Isles would be:

- colder in summer and winter as further from the Equator and the Sun's rays are more dispersed
- colder as more mountainous (at higher altitude)
- colder as higher wind speed as at higher altitude
- wetter as more relief rainfall as at higher altitude and prevailing wind comes from the south-west so has evaporated moisture over the Atlantic
- affected by the Gulf Stream so temperatures will be slightly warmed in winter but effect of latitude will still make temperatures colder than in south. (6)

15 Water cycle (6)

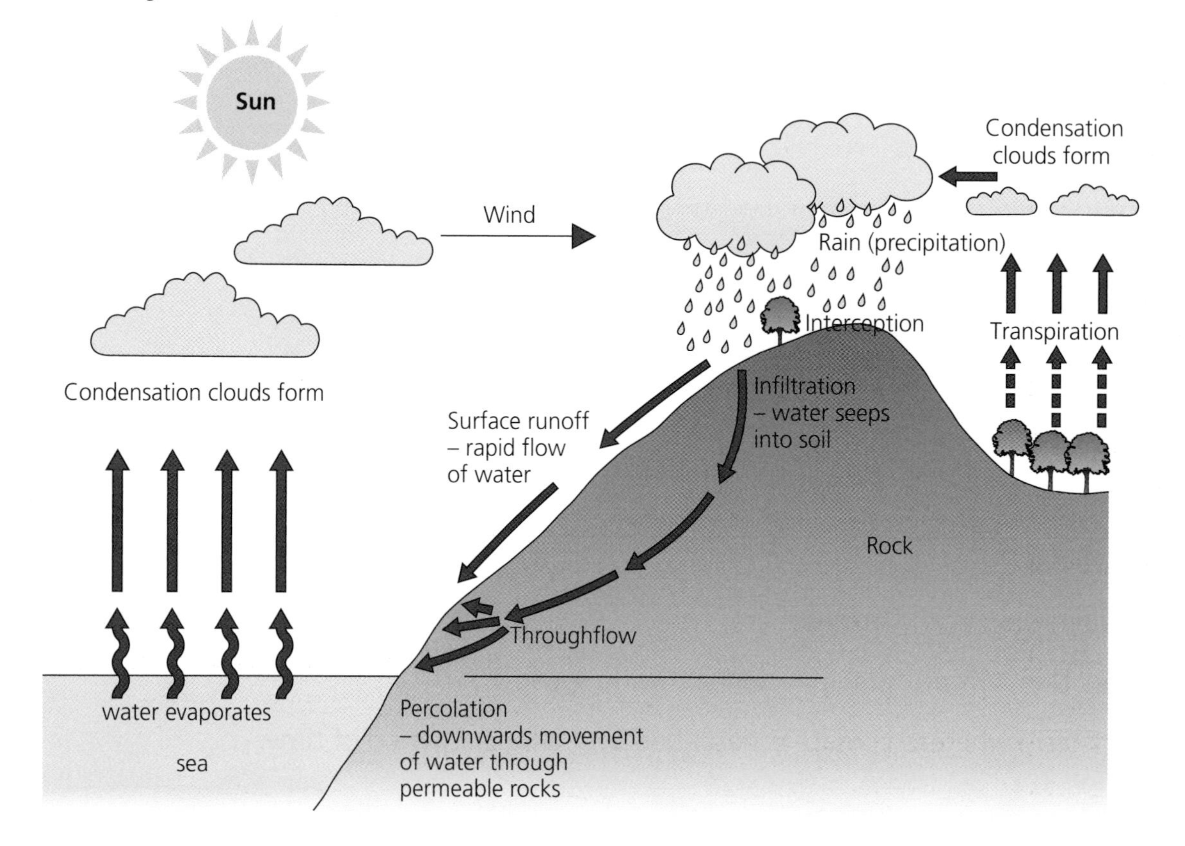

16 Evaporation occurs when water turns to vapour due to cooling. (1)

Precipitation is anything wet that falls from the sky, for example rain, snow, sleet and hail. (1)

17 *Two definitions only and a maximum of three marks for each.*

Mountain

- Higher altitude causes lower temperatures.
- Higher altitude causes more relief rainfall.
- Higher altitude causes higher wind speed.
- Lower temperature causes more snow. (3)

Aspect

- A slope with a southerly aspect in the northern hemisphere is warmer than one with a northerly aspect, due to the tilt of the Earth.
- A slope with an easterly aspect has more Sun in the morning and a slope with a westerly aspect has more Sun in the evening. (3)

Coast

- Places near the coast are warmer in winter than places inland.
- Places near the coast are colder in summer than places inland.
- Places near the coast have higher wind speeds due to lack of shelter. (3)

Forest

- Shade reduces sunlight and temperature.
- Shelter reduces wind speed but increases temperature. (3)

Town

- Dark tarmac surfaces increase temperature as they absorb more sunlight.
- Buildings reduce wind speed and therefore temperature.
- Buildings release heat from central heating systems and as the bricks absorb heat in the day. (3)

18 (a) Any two of:

- The temperature is not as hot as at the Equator but not as cold as at the North Pole.
- The Sun's rays heat the British Isles at an angle and therefore the rays are dispersed and do not heat up the Earth's surface as much as they do at the Equator where they are concentrated on a small area.
- The rays travel further through the atmosphere and much heat is lost to dust particles. (2)

(b) Any two of:

- increases temperatures
- in the west of the British Isles
- in winter. (2)

19

Letter	Climate description
A	mild summers, mild winters and wet
B	mild summers, very cold winters and dry
C	warm summers, mild winters and wet
D	warm summers, cold winters and dry

(4)

20 (a) Site 3 (1)

Any two of:

- It has no shelter.
- It has no shade.
- It does not have a tarmac surface which absorbs heat.
- It has easy access from the school. (2)

(b) Site 2 (1)

Any two of:

- It has a tarmac surface which absorbs heat.
- It has no shade.
- It has shelter from the prevailing wind. (2)

21 Any three of:

- transpiration – the release of water vapour from the leaves of vegetation
- evaporation – water turning to vapour due to cooling
- condensation – water vapour turning to liquid due to cooling
- infiltration – vertical movement of water through soil
- surface run-off – movement of water over the surface of soil
- throughflow – horizontal movement of water through soil
- precipitation – water falling from the sky in the form of snow, rain, hail, etc. (3)

22 (6)

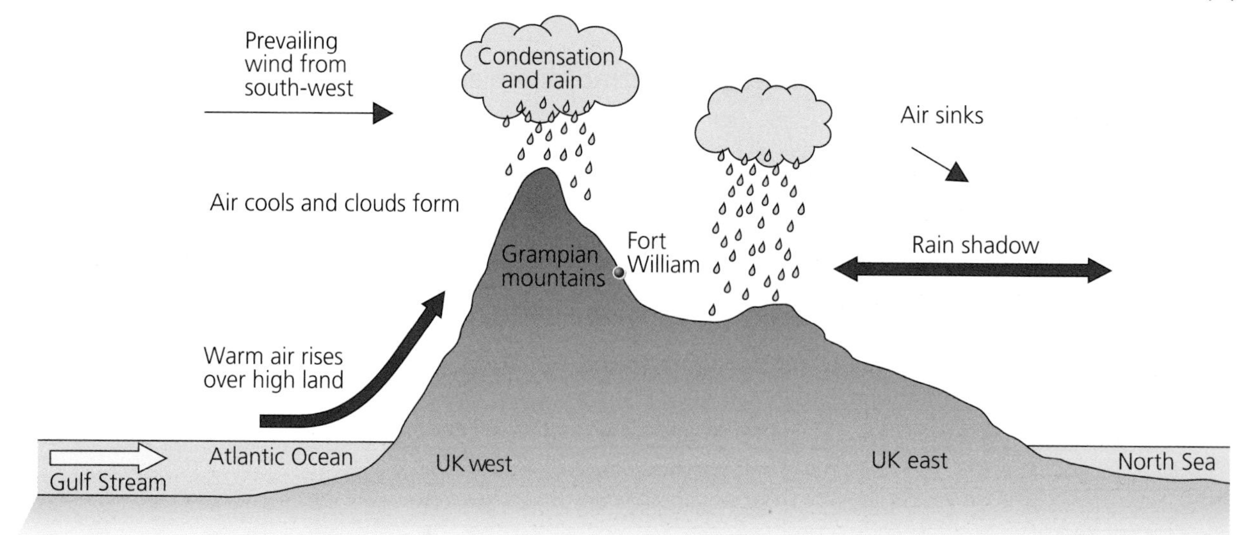

23 (a) Fort William (1)

Any one of:

- It is in a mountainous area so has relief rainfall.
- It is in the west so the prevailing wind has carried moist air from the Atlantic to the hills around Fort William.
- It is in the west of the country. (1)

(b) 1975 mm (1)

(c) Norwich (1)

Any one of:

- It is in the south so nearer the Equator.
- It is at low altitude. (1)

(d) Plymouth (1)

Any one of:

- It is surrounded by sea.
- It is in the south so it is nearer the Equator.
- It is affected by the Gulf Stream. (1)

(e) Fort William (1)

Any one of:

- It is at high altitude.
- It is far away from the Equator. (1)

(f) Norwich (1)

24 (a) Any two of:

- It is in the north of the British Isles so far away from the Equator.
- It is in the Grampian mountains at high altitude, so it is cold.
- It is in the north-east rather than the north-west so not affected by the Gulf Stream.
- It is in the mountains so there is likely to be precipitation (relief). (2)

(b) Gulf Stream or North Atlantic Drift or direction of ocean current (1)

(c) The isotherms would show temperature increasing from north to south. This is because latitude is the main influence on the temperature in the summer as the Gulf Stream only affects the British Isles in winter. (2)

25 Any three of:

- clouds
- soil
- river
- sea
- glacier. (3)

26 Any two of:

- In the summer places near the sea are cooler than places inland as the sea takes a long time to heat up.
- In the winter places near the sea are warmer than places inland as the sea retains its heat in winter.
- Places on the west coast in winter are also warmed by the Gulf Stream. (2)

27 Any two of:

- The Sun's rays shine directly on the Equator.
- The Sun's rays hit places at higher latitudes at an angle and therefore are dispersed.
- The Sun's rays have to travel further to get to places further from the Equator and therefore some heat is lost to dust in the atmosphere. (2)

28 *One mark for correct factor and one mark for explanation:*

- Altitude – the higher the altitude the lower the temperature.
- Wind speed – the higher the wind speed the lower the temperature. (2)

29 Any two of:

- Darker surfaces absorb more heat.
- More buildings reduce wind speed.
- Buildings release warmth from central heating.
- Bricks retain heat at night. (2)

30

Type of rainfall	Place
relief rainfall	Ben Nevis
frontal rainfall	the whole of the UK
convectional rainfall	London

(3)

3 Rivers and coasts

1 (a) a method of transportation (1)

(b) all the material that a river carries (1)

(c) load scraping against the bed and bank (1)

(d) the uninterrupted distance a wave travels (1)

(e) a feature of erosion (1)

(f) the outside bend of a meander (1)

(g) fine load being carried in the flow (1)

(h) on the outside of a bend in a river (1)

(i) on the river beach (1)

(j) spit (1)

2 *Three definitions only and any two points for each:*

Hydraulic action

- a type of erosion
- where the force of the water hits against the bed and the banks of a river or cliff forcing air into cracks in the bed and banks or cliff (2)

Saltation

- a type of transportation
- where medium-sized pebbles, small pebbles or grains of sand are bounced along the river bed hitting each other to push each other along (2)

Corrosion

- a type of erosion
- where the acid in the sea or river water attacks the river bed and banks (2)

Deposition

- the action of load being dropped
- by slow-flowing water (2)

3 (a) Weathering is the breakdown of rock by the weather, plants and animals. (1)

Erosion is the breakdown and removal of rock by rivers, sea and ice. (1)

(b) Any one of each:

Weathering:

- freeze-thaw (frost shattering)
- onion-skin (exfoliation)
- chemical
- biological (1)

Erosion:

- hydraulic action
- abrasion
- attrition
- corrosion (1)

4

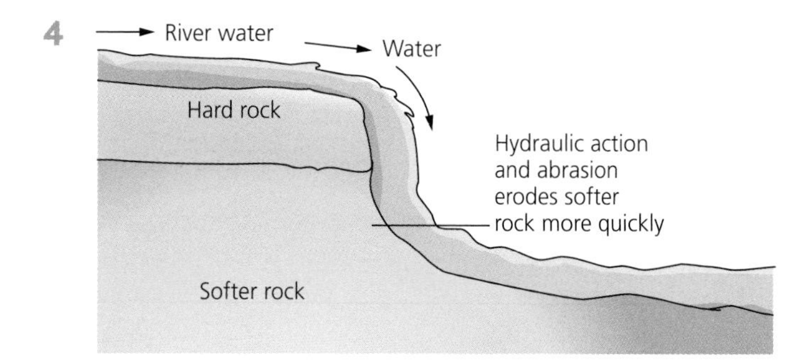

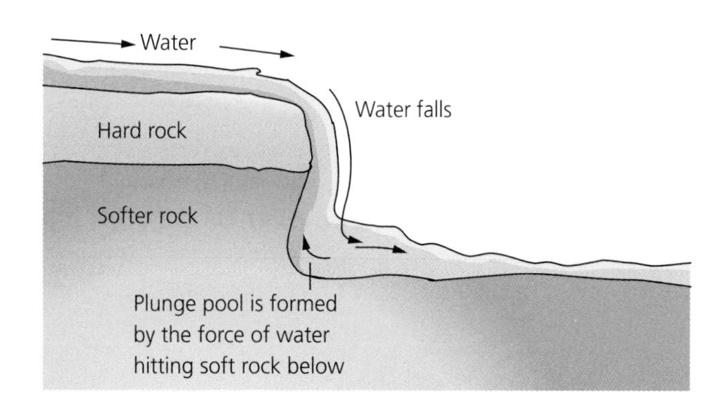

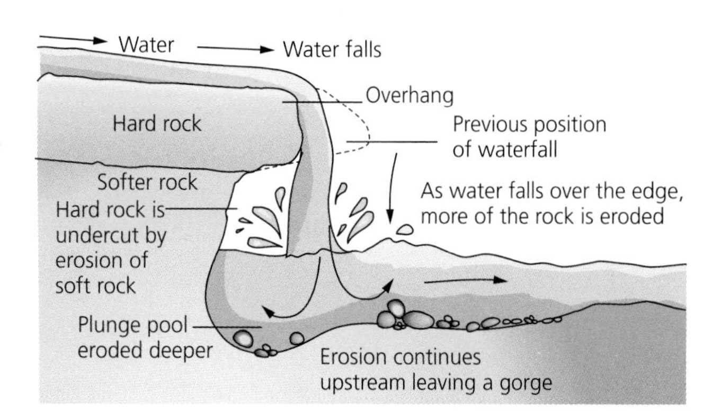

(5)

5

undercutting
river cliff
river beach
shallow water
fast flowing water

(5)

6

River landform	Grid square	Process
meander (1)	many different grid references (1)	both (1)
flood plain (1)	2985, 2785 or other (1)	deposition (1)

7 Any three of:

- Groynes have been created.
- These stop longshore drift which is a natural physical process.
- The swash and backwash will be interrupted.
- The sediment/beach will build up on one side of the groyne and the beach will become narrower on the other side of the groyne. (3)

8

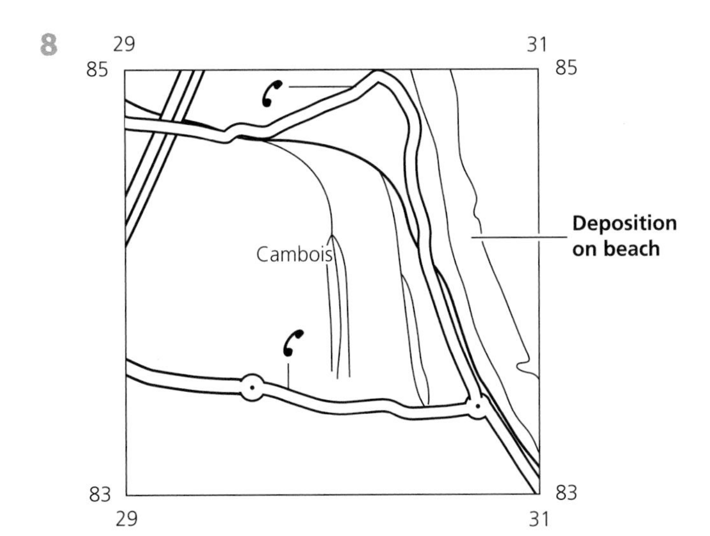

(2)

9

Type of weathering	
onion-skin weathering (exfoliation)	desert (1)
freeze-thaw weathering	mountain (1)
biological weathering	chalk hills (1)
chemical weathering	limestone (1)

(4)

10 (a) source (1)

(b) confluence (1)

(c) drainage basin (1)

(d) floodplain (1)

(e) tributary (1)

(f) impermeable (1)

(g) stump (1)

(h) exfoliation (1)

(i) weathering (1)

(j) swash (1)

(k) scree (1)

11 (a) whinstone (1)

(b) Any two of:

- It has been undercut.
- It has been left as an overhang.
- It has not been eroded as much as the shale. (2)

(c) Any three of:

The plunge pool is deeper because of:

- hydraulic action from the water as it falls into the plunge pool
- the velocity of the water as it falls over the overhang causing an increase in hydraulic action
- abrasion from the collapsing overhang
- abrasion from the loose rock in the plunge pool swirling around. (3)

(d) In 500 years' time (1)

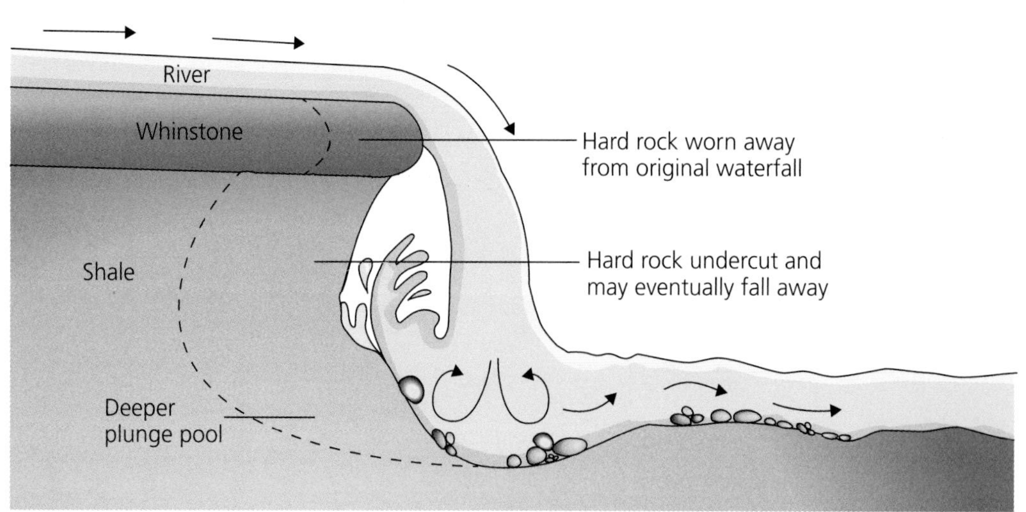

(e) Any three of:

- The shale will be eroded further by hydraulic action, abrasion and corrosion.
- The whinstone will be undercut further.
- The whinstone will eventually collapse due to a lack of support.
- The waterfall will retreat towards the source.
- This process will keep on repeating itself.
- The valley sides will be left as a steep-sided gorge. (3)

12 (a) Where, any one of:

- on the river beach
- on the inside bend of a meander
- at the mouth of the river
- at an estuary or delta. (1)

Why, any one of:

- The velocity of the water decreases.
- The water slows down. (1)

(b) Any one of:

- river beach
- floodplain
- estuary
- delta
- ox-bow lake. (1)

13 (a) (4)

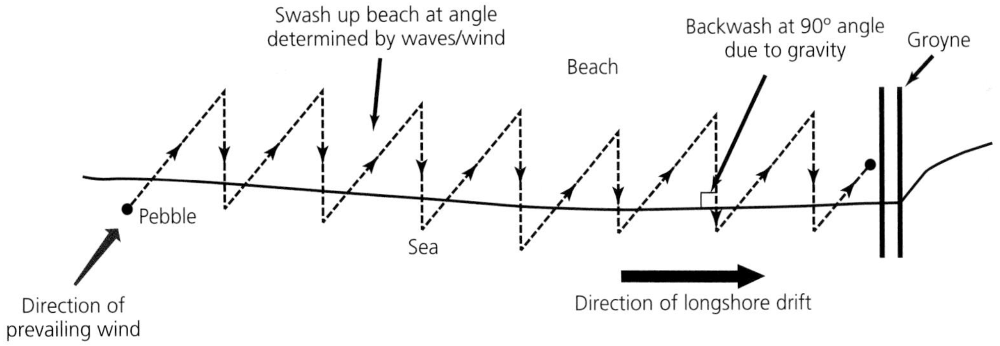

(b) Any one of:

- spit
- beach. (1)

(c) Diagram to show formation of spit or beach (6)

Formation of spit

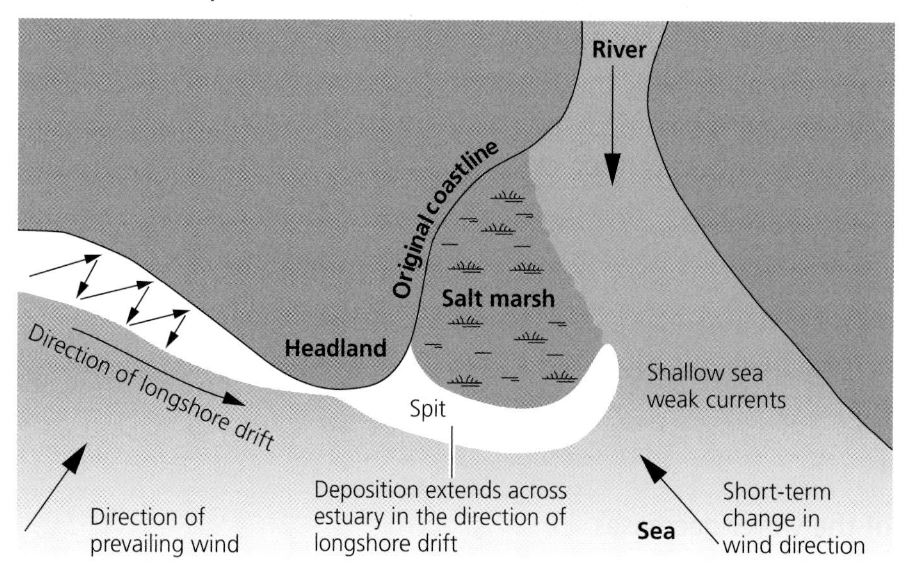

Formation of beach

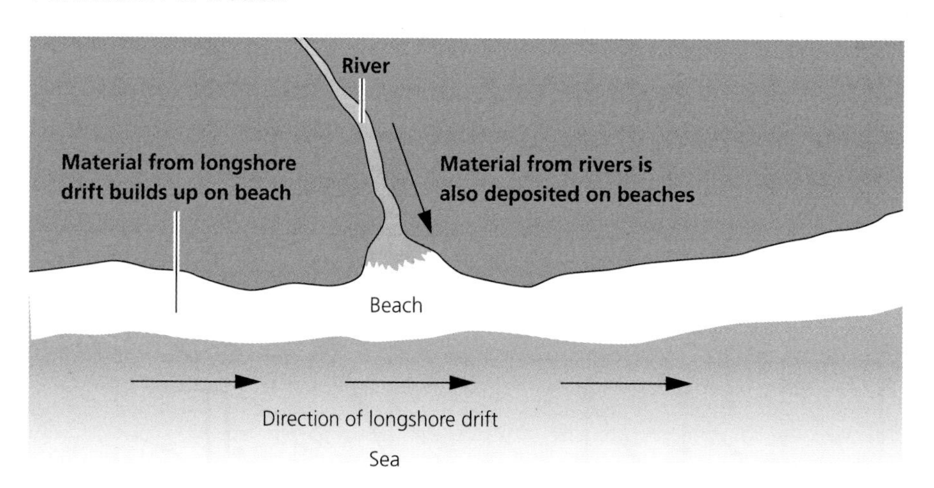

14 (a) biological weathering (1)

(b) freeze-thaw weathering or frost shattering (1)

(c) Any three of:

- Rain water enters joints/cracks/faults.
- Water freezes.
- The volume of the ice puts outward pressure on the fault.
- The rock breaks up.
- Scree is formed.

(3)

15

Erosion	Deposition
arch	beach
stump	spit
waterfall	floodplain

(6)

16

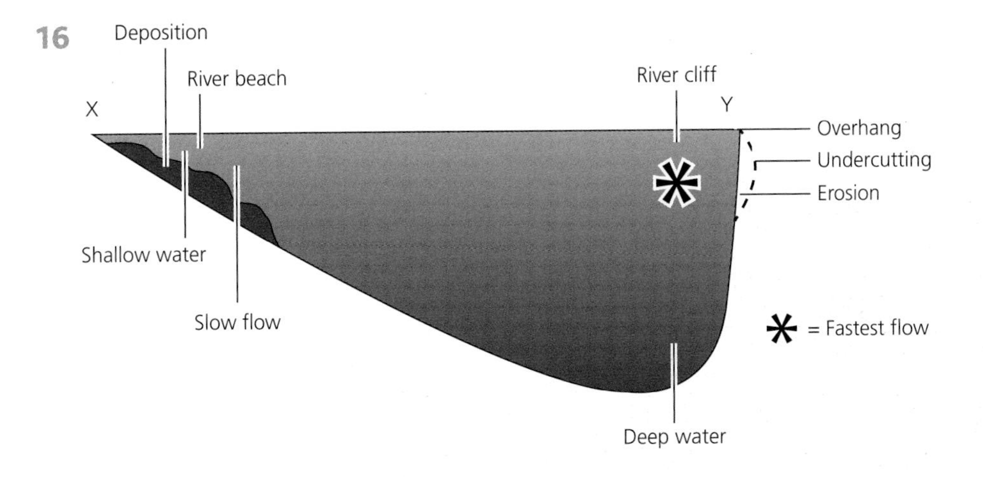

(12)

17

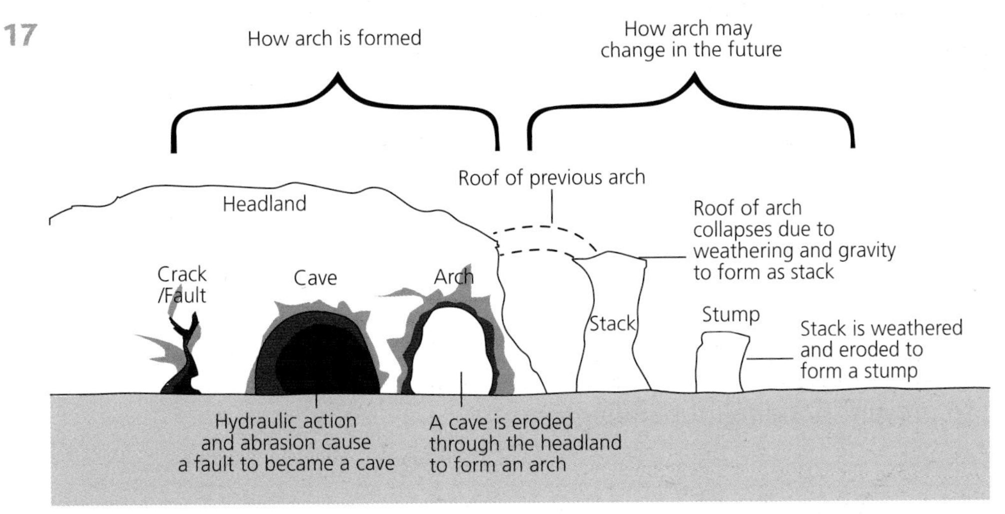

(6)

18 (a) cause (1)

(b) cause (1)

(c) effect (1)

(d) cause (1)

(e) cause (1)

(f) effect (1)

(g) cause (1)

19 (a) *Answers depend on the case study chosen. One mark can be given for each point and two marks if the point is developed, to a maximum of four marks.*

- heavy rain over a prolonged period of time
- saturated soil, leading to increased run-off and decreased infiltration
- impermeable soil
- steep slopes
- urbanisation on flood plain
- deforestation
- straightening of channel

(4)

(b) (i) *One mark can be given for each point and two marks if the point is developed, to a maximum of three marks.*

- Vegetation may be destroyed.
- Wildlife can be killed or disorientated.
- The built environment can be damaged or destroyed. (3)

(ii) *One mark can be given for each point and two marks if the point is developed, to a maximum of three marks.*

- People could drown.
- People could have to evacuate.
- Communications (roads, bridges, telephone cables, etc.) could be destroyed.
- Services (hospitals, schools, etc.) could close.
- Diseases such as dysentery and cholera could spread through people drinking dirty water. (3)

(c) *One mark can be given for each point and two marks if the point is developed, to a maximum of four marks.*

- construction of dams
- construction of levees and dykes to contain water
- straightening of meanders to remove flood water quickly
- afforestation to increase transportation and infiltration
- sandbagging to prevent flooding of buildings (4)

20 (a) erosion (1)

(b) deposition (1)

(c) impermeable clay soil (1)

(d) slip-off slope (1)

(e) physical (1)

21 (a) (i) lower course of the river (near the mouth) (1)

(ii) upper course of the river (near the source) (1)

(iii) middle course (1)

(b) Any two of:

- Particles of load collide and knock pieces off each other (attrition).
- Particles of load leapfrog and roll along the river bed causing them to break up.
- Acid in the water dissolves the load (corrosion). (2)

22 (a) Any three of:

- The slow velocity will cause deposition to occur.
- The load will be deposited.
- The sediment on the river beach will build up. (3)

(b) Any three of:

- Hydraulic action, abrasion and attrition will erode the river cliff.
- The bank will be undercut.
- The overhang may eventually fall into the river.
- The river bed will be deepened.

(3)

23

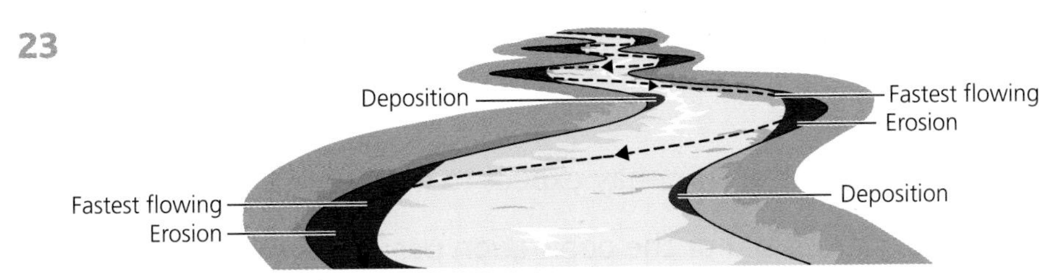

(3)

4 Population and settlement

1 *Two definitions only and one mark for each:*

Life expectancy – the average period that a person may expect to live. (1)

Birth rate – number of live births per thousand of the population per year. (1)

Migration – the movement of people from one place to another with the intention of settling. (1)

Natural increase – the birth rate minus the death rate, which tells how quickly a population is growing. (1)

2 A push factor is a reason why people may leave one place and migrate to another whereas a pull factor is a positive point attracting people to the new location. (2)

Examples of push factors (one required for one mark):

- lack of jobs
- famine
- natural disaster
- war
- poor education facilities
- high crime rates
- expensive houses (1)

Examples of pull factors (one required for one mark):

- political stability
- job opportunities
- good health care system
- good schools
- low crime rates
- cheap houses/big gardens
- space (1)

3 *One mark for each point up to a maximum of three marks.*

- expensive houses
- lack of space/small gardens

- pollution
- high crime rates
- noise (3)

4 *One mark for each point up to a maximum of three marks.*

- job opportunities
- perceived better standard of living
- better schools
- better health care
- to be with friends or relatives (3)

5 (a) Check current population of the world (1)

(b) 262 people per square kilometre (as of 2014) (1)

(c) *One mark for each point up to a maximum of three marks.*

- more job opportunities
- warmer
- lower rainfall
- flatter relief
- better transport links (3)

6 Amazon Rainforest – 0.2 people per square kilometre (1)

Russia – 8.4 people per square kilometre (1)

Italy – 199 people per square kilometre (1)

7 (a) *One mark for each correct statement:*

The most densely populated areas are India, Bangladesh, Korea and Japan.

Europe, China, Indonesia and parts of Africa are also densely populated.

The USA, South America and the Middle East are less densely populated.

Canada, Russia and Australia are sparsely populated. (4)

(b) The places with a high population density have high rates of natural increase or immigration or both. (1)

The places with a high population density may have favourable physical conditions and the places with sparse populations may not. These favourable physical conditions could be:

- temperate climates
- fertile soil
- flat land
- resources such as oil, coal, wood or fish. (2)

The places with a high population density may have favourable human conditions and the places with sparse population may not. These favourable human conditions could be:

- job opportunities
- transport links
- stable governments. (2)

8 (a) (5)

	Birth rate	Death rate	Rate of natural increase
France	12	9	**3**
Japan	8	9	**−1**
Ethiopia	**42**	11	**31**
Afghanistan	39	**14**	25

(b) The birth rate is higher in the less developed countries. (1)

The death rate is higher in the less developed countries. (1)

The rate of natural increase is higher in the less developed countries. (1)

9 *Answer depends on the country chosen.* (5)

10 (a) greenfield site (1)

(b) by-pass (1)

(c) semi-detached (1)

(d) dispersed (1)

11 *Answers depend on the case study chosen.*

(a) (3)

(b) (2)

(c) (6)

12 *One mark for each point and an extra mark if the point is developed, to a maximum of four marks.*

- Environmentalists would be against this development as it would cause the destruction of habitats.
- Residents in the suburbs may be annoyed as the supermarket may bring extra cars to the roads causing traffic jams.
- Residents may be annoyed by the visual pollution of the supermarket and its car park.
- More lorries may cause noise on the roads as they bring deliveries to the supermarkets.
- The caravan and camp site and the picnic site could lose their serenity, become noisy and no longer have attractive scenery. (4)

13 *One mark for each point up to the maximum given in brackets.*

(a) • It is sited on a river so has very good access by boat.
• It is sited on the coastline and has a harbour.
• It is on a relatively flat site. (2)

(b) • It is on a relatively flat site.
• It is sited on the River Wansbeck which gives good access by boat.
• It is surrounded by flat and fertile agricultural land on a floodplain.
• It is situated near woodland which could have been used for fuel and building materials. (3)

14 *One mark for each point up to a maximum of two marks.*

- It is in competition with Pegswood and Morpeth which both have train stations.
- It is only accessible by B roads rather than A roads.
- It does not provide many job opportunities.
- There are many lakes/ponds to the south-west of the settlement which limits development in this direction. (2)

15 Longhirst = linear

Ulgham = nucleated

Stannington Station = linear

Hebron = nucleated (4)

16 *One mark for each point up to the maximum given in brackets.*

(a) It is situated:

- along a coastline
- in a valley floor
- along a road
- along a river. (2)

(b) • It is situated in a mountainous area and there is very little flat space to use for building.
• It is in an agricultural area and each building is surrounded by its farmland. (1)

17 *One mark for each point up to the maximum given in brackets.*

(a) Morpeth:

- has many more shops and services than Longhirst
- probably has some large supermarkets and specialist shops whereas Longhirst probably has no shops or just one convenience store
- has six places of worship and Longhirst only has one
- has four schools and Longhirst has none
- has a hospital but Longhirst does not
- has many tourist facilities but Longhirst does not
- has a train and bus station but Longhirst does not. (3)

(b) Morpeth:

- is a town and Longhirst is a village
- is a higher order settlement
- is higher on the settlement hierarchy
- has a larger population
- is more likely to get more tourists as it is a larger settlement. (2)

18 Settlement hierarchy (8)

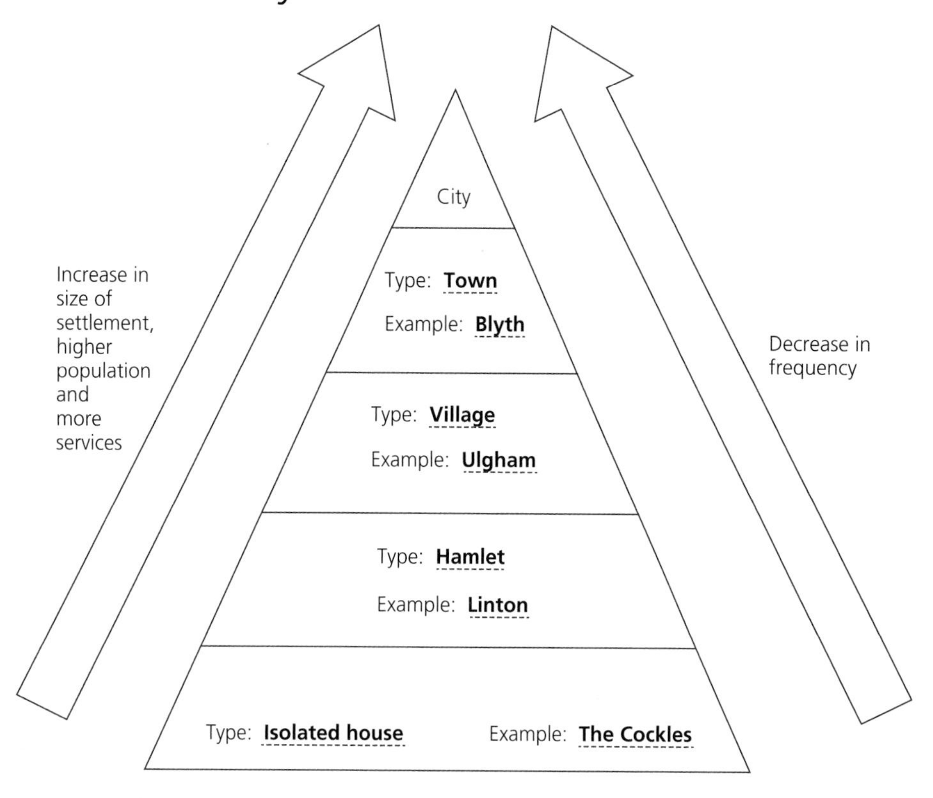

19 (a)
- It is sited on a river which is useful for transportation, fishing, washing and drinking.
- There is some mixed woodland to the north which would be useful for providing building materials and fuel.
- It is a flat site which is easy to build on.
- It is on a floodplain which provides fertile soil for agriculture.
- It is near the coastline which is useful for fishing and transportation. (3)

(b) Any three of the following:

- hospitals
- schools
- places of worship
- leisure centre
- golf course (3)

(c) Any three of the following:

- nature reserves
- open gardens
- museum
- information centre (3)

20 *One mark for each point up to a maximum of two marks.*

- The CBD will have many more shops than the suburbs.
- The CBD will have many more specialist shops.
- Suburbs will have more convenience stores which serve the immediate area.
- The suburbs may have a retail park with large superstores. (2)

21 A = semi-detached

B = terraced

C = detached (3)

22 Suburbs (1)

5 Transport and industry

1 *One mark for each point and an extra mark if the point is developed, to a maximum of three marks.*

- To enable the labour to get to work
- To get the raw materials to the factory
- To get the cakes and biscuits to the market (3)

2 *One mark for each point, to a maximum of two marks.*

- It is quicker.
- It relieves congestion on roads.
- It is safer. (2)

3 Many people can travel in one vehicle. (1)

4 *One mark for each point, to a maximum of two marks.*

- It is cheaper. (1)
- It is more environmentally friendly. (1)
- Less carbon dioxide is emitted. (1)

5 Air. The flowers perish quickly so need to get to the destination quickly. (2)

6 *Answers depend on the case study chosen. Reward answers that have used specific names of places, plants or animals.* (4)

7 *Answers depend on the case study chosen. Reward answers that have used specific names of places.* (4)

8 *Answers depend on the case study chosen.* (4)

9 Pegswood (1)

One mark for each point, to a maximum of two marks.

- Pegswood has a train station.
- Pegswood is next to the A197 which is a main road.
- Ulgham is only accessed by the B1337 and minor roads/lanes. (2)

10 *One mark for each point, to a maximum of three marks.*

- Morpeth has a train station.
- The A197, A192 and A196 all lead into Morpeth.
- Morpeth is also accessible by the River Wansbeck.
- Morpeth is also served by many minor roads. (3)

11 (a) sustainable development (1)

(b) eco-tourism (1)

12 (a) Looking after resources in a sustainable way for the future (1)

(b) To seek and to use a natural resource for human benefit (1)

13 (a) Looking after resources in a sustainable way for the future (1)

(b) *Reward any suitable answer.* (1)

14 *One mark for each point and an extra mark if the point is developed, to a maximum of three marks.*

- There may be noise from lorries on the A189 and minor roads carrying raw materials and finished product.
- A greenfield site may have been used to build the factory, in which case habitats would have been destroyed.
- The factory itself may be an eyesore, causing visual pollution especially for tourists at the coast or in the Queen Elizabeth Country Park and for inhabitants of Lynemouth.
- Some emissions could end up in Woodhorn Burn or in the sea. This could affect wildlife. (3)

15 (a) primary industry (1)

(b) tertiary industry (1)

(c) service industry (1)

(d) secondary industry (1)

(e) secondary industry (1)

(f) is service industry (1)

(g) in large metal boxes (1)

16 (a)

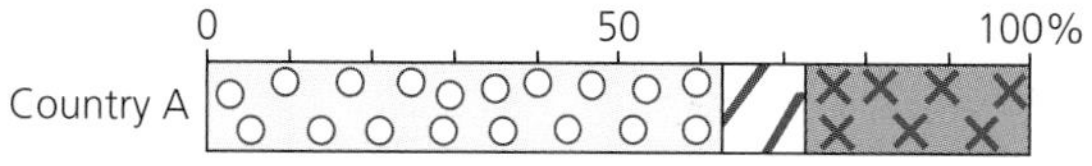

(2)

(b) A: less developed country

B: more developed country (2)

(c) Any three of:

- Country A has 21 times more people working in primary industry.
- Country B has more than twice as many people working in tertiary industry.
- Country A has no one working in quaternary industry whereas Country B has some.
- Country A has less than half as many people working in secondary industry than Country B. (3)

(d) *One mark for each point and an extra mark if the point is developed, to a maximum of three marks.*

- Many people in less developed countries work in agriculture (perhaps subsistence farming) as their priority is to feed themselves and their family. Their farming is not mechanised.
- In less developed countries people often work in the mining industry but often for a company from a rich country, and the less developed country cannot afford to exploit the minerals themselves.
- The developed country has a small percentage in primary industry due to mechanisation and the fact that it is often more economical to import foodstuffs from abroad.
- The developed country has a higher percentage in tertiary industries as the country can afford to employ more teachers, doctors, etc. Also the population has greater wealth to spend on eating out, shopping and leisure activities.
- The developed country has more people in quaternary industry as it has the technology and money to research and develop. (3)

17 (a) an industry involved in the extraction or growing of raw materials, for example agriculture (1)

(b) an industry involved in research and development, for example microbiology (1)

(c) a newly industrialised country (1)

18 (a) *Three location factors only and a maximum of two marks for each. One mark for each point and an extra mark if the point is developed.*

- Flat relief and dry site: it is easier to build on flat land than a slope, and a dry site away from marsh land is necessary. (2)
- Government grants: the government can give grants (let companies off rates) if the factory is located in an area of high unemployment, as this will create wealth in poor areas. (2)
- Raw materials: it is important that the factory manager considers where the raw materials are coming from as they will cost money to transport. (2)
- Transport: the raw materials have to be transported to the factory and the finished product has to be taken away from the factory. Transport links are also important for the labour force. (2)

- Settlement: it is useful to have a settlement near to the factory so that the labour supply is nearby. The settlement could also be the market. (2)
- Power station: this is no longer an important factor in the location of a factory because power is ubiquitous. (2)

(b) Being near a power station is not important in modern times because power is ubiquitous. (2)

Being near raw materials is not as important because transport is now more efficient. (2)

19 (a) Any three of:

- 2985 is on the edge of the town and 2786 is in the centre of the town.
- 2985 is surrounded by an open space area and 2786 is surrounded by a residential area.
- 2985 is nearer to the coast than 2786.
- 2985 is next to the A189 and 2786 is next to the A196. (3)

(b) *One mark for each point and an extra mark if the point is developed, to a maximum of four marks.*

- 2985
- outside of the town so any noise or smell pollution does not affect the residents
- outside of the town so the price of land is cheaper
- outside of the town so there is space for expansion and car parks
- near to the town for labour supply
- near to the town as a possible market for the goods
- near to a main road for the employees to get to work, for the raw materials to be imported and the finished product to be exported
- flat land (4)

20

	Example	Grid reference
primary	farm/quarry	many grid references (2)
secondary	factory	many grid references (2)
tertiary	museum, hospital, school, etc	many grid references (2)

21 (a) primary (1)

(b) *One mark for each point and an extra mark if the point is developed, to a maximum of four marks.*

- disturbance by noise of machinery, etc.
- noisy and dusty lorries through Morpeth
- congestion on roads from lorries
- eyesore (visual pollution) for residents
- tourists are put off coming due to visual pollution
- destruction of wildlife habitats (4)

(c) *One mark for each point and an extra mark if the point is developed, to a maximum of four marks.*

- Loss of jobs in surrounding area.
- Services and shops in Morpeth will suffer as residents have less money to spend.
- Crime may increase as unemployment leads to poverty.
- Emigration may occur as people leave to find jobs elsewhere.

- The social structure of the area may change resulting in more elderly people.
- Quarries may be left disused as a danger and an eyesore.
- The environment may improve as the quarries are made into lakes or infilled to create new wildlife habitats. (4)

(d) Tourism (1)

Evidence could be information centres, museums, visitor centres or camp and caravan sites. (1)

22

Primary	Secondary	Tertiary	Quaternary
forestry, arable farming	cement works	law, tourism, entertainment	R&D
(2)	(1)	(3)	(1)

23 *Answers depend on the case study chosen.*

(a) (2)

(b) (3)

(c) (3)

24 (a) Any three of:

- Fewer people work in primary industry.
- Fewer people work in secondary industry.
- More people work in tertiary industry.
- People now work in quaternary industry. (3)

(b) *One mark for each point and an extra mark if the point is developed, to a maximum of three marks.*

- Primary has declined as it is cheaper to import some agricultural products and due to mechanisation on farms. Mining has also declined due to cheaper coal being imported from countries such as Poland and cleaner, more efficient forms of energy being used. There has been mechanisation in mining.
- Secondary has declined as many companies are locating their manufacturing in less economically developed countries as the cost of labour is much cheaper. Again, mechanisation has meant that fewer people are needed in factories.
- Tertiary has increased as people become wealthier and have a greater disposable income to spend on leisure. Also more taxes are collected so public services can be improved.
- Quaternary has come about as scientific advance has allowed this kind of job to exist. (3)

25 Any four of:

- 5 mins from the M4 – good for getting to meetings, employees getting to work, importing raw materials and exporting finished product
- 40 mins to Heathrow – good for getting to meetings, importing raw materials and exporting finished product

- Landscaped gardens – pleasant environment, which could attract employees
- Car parking for 500 cars – convenient for employees and people coming for meetings
- Wi-Fi installed – shows that the offices have the latest technology
- Many big names already here – companies now locating here get prestige from being near a big name and could possibly share some facilities (4)

26 *Answers depend on the case study chosen.*

(a) (6)

(b) (2)

(c) (4)

27 *Answers depend on the case study chosen.*

Could include:

- wages
- working hours
- holiday pay. (3)

28 **Primary** industry involves the growing or extracting of raw material on or from the Earth. In less economically developed countries a **large** proportion of the population works in this sector. Many people are subsistence **farmers**. Once a country begins to **develop**, the number of people working in **secondary** industries will increase. However, most of these people will be working in factories owned by companies whose headquarters are in **developed countries**.

In more developed countries the majority of people work in **tertiary** industries, especially in **urban** areas. A very small proportion of the population is involved in **quaternary** industries which are concerned with research and development. The employment structure of a country or region changes over **time**. Two hundred years ago far fewer people in the UK worked in **tertiary** industries and far more worked in **primary** industries as farmers, miners or **fishermen**. (13)

29 *Answers depend on the case study chosen.* (3)

6 Location knowledge

1 (a) See Map 1 later in the chapter (Madrid, Berlin, Athens, Rome) (4)

(b) See Map 1 (pupils should have marked two of the four countries: Spain, Germany, Greece, Italy) (2)

(c) Alps (1)

(d) See Map 1 (English Channel) (1)

(e) River Rhine (1)

(f) See Map 1 (Prime Meridian, Pyrenees) (2)

2 (a) See Map 2 later in the chapter (New York) (1)

(b) See Map 2 (Mexico City) (1)

(c) See Map 2 (River Amazon) (1)

(d) See Map 2 (Brazil and Argentina) (2)

(e) Brasilia (1)

(f) See Map 2 (Andes) (1)

(g) See Map 2 (Peru and Colombia) (2)

(h) Pacific Ocean (1)

(i) Rocky Mountains (1)

(j) See Map 2 (Canada, Ottawa and Vancouver) (3)

(k) See Map 2 (Mississippi River) (1)

3 (a) Kenya (1)

(b) Nairobi (1)

(c) France (1)

(d) Paris (1)

(e) River Nile (1)

(f) Cairo (1)

(g) See Map 3 later in the chapter (Ghana and Nigeria) (2)

(h) D = Atlantic Ocean (1)

E = Indian Ocean (1)

(i) See Map 3 (Sahara Desert) (1)

(j) Back (1)

4 (a) 1 = Mississippi River (1)

2 = River Amazon (1)

3 = River Nile (1)

4 = River Rhine (1)

5 = Yangtze River (1)

(b) (i) Egypt (1)

(ii) Cairo (1)

(c) Alps (1)

(d) See Map 4 later in the chapter (New Delhi) (1)

(e) Back (1)

(f) See Map 4 (Norway and Copenhagen) (2)

5 (a) See Map 5 later in the chapter (Sydney) (1)

(b) United States of America (1)

(c) Morning (1)

(d) Summer (1)

(e) B = Russia (1)

C = Germany (1)

D = Brazil (1)

(f) Pyrenees (1)

(g) (i) F = Equator (1)

(ii) G = Indonesia (1)

(iii) H = Tropic of Capricorn (1)

(h) Athens (1)

6 (a) See Map 6 later in the chapter (Oslo) (1)

(b) Norway (1)

(c) North Sea (1)

(d) See Map 6 (Saudi Arabia) (1)

(e) Tropic of Cancer (1)

(f) B (1)

(g) A (1)

(h) Tropic of Capricorn (1)

(i) Atlantic Ocean (1)

7 (a) A = Belfast (1)

B = Cardiff (1)

C = Birmingham (1)

D = Madrid (1)

(b) E = Poland (1)

F = The Netherlands (1)

(c) See Map 7 later in the chapter (Alps, Pyrenees, Grampian mountains) (3)

(d) (i) Dublin (1)

(ii) Berlin (1)

(iii) Lisbon (1)

(iv) Bern (1)

(v) Warsaw (1)

(e) See Map 7 (Arctic Ocean, English Channel, River Clyde, River Severn) (4)

(f) England, Wales, Scotland and Northern Ireland (4)

8 (a) Bangladesh (1)

(b) See Map 8 later in the chapter (Himalayas) (1)

(c) See Map 8 (Pakistan) (1)

(d) Yangtze River (1)

(e) See Map 8 (Indonesia) (1)

(f) Tokyo (1)

(g) See Map 8 (India) (1)

(h) C = Iraq – Baghdad (2)

D = Iran – Tehran (2)

(i) Arctic Circle (1)

66.6° N (1)

(j) December (1)

Map 1

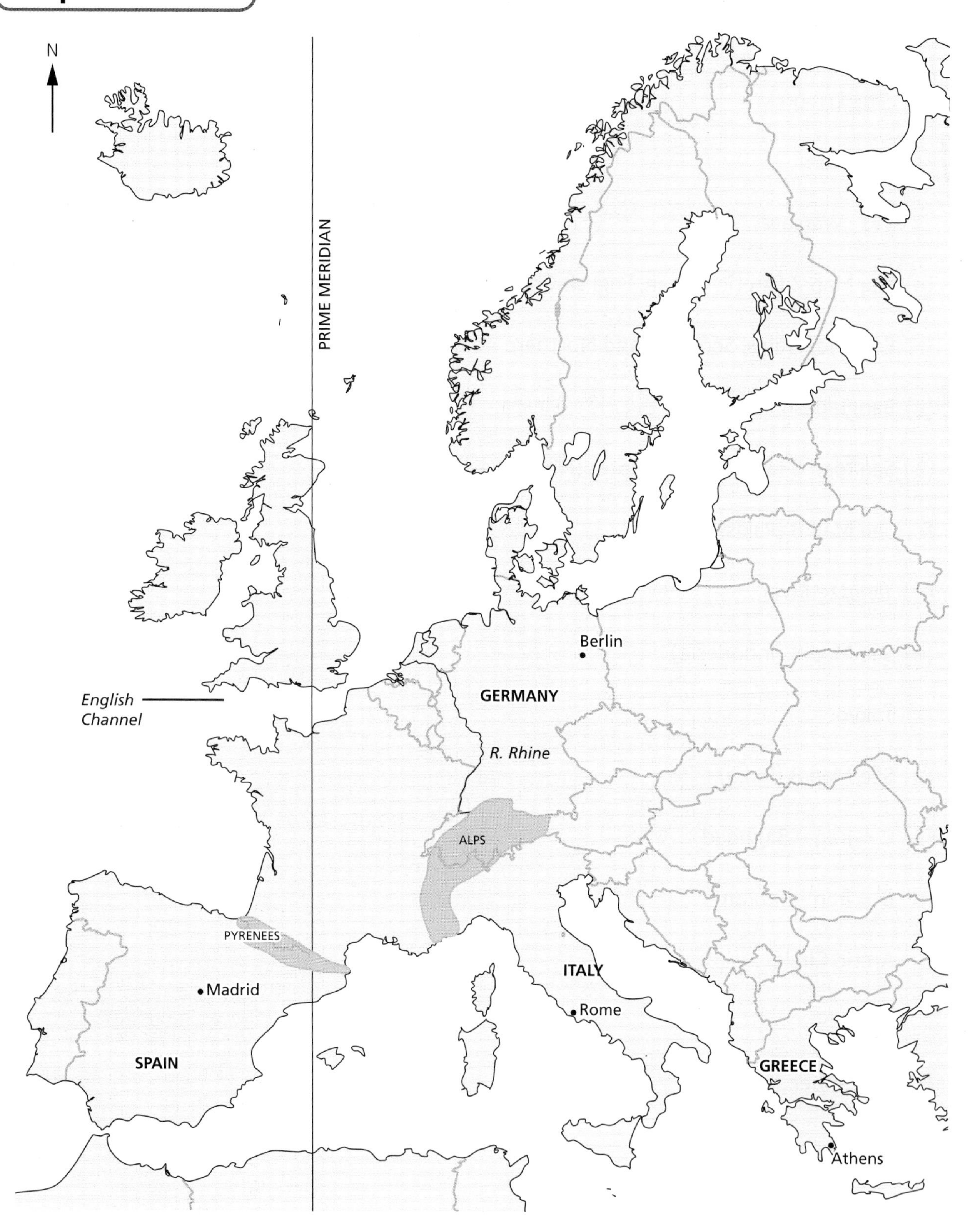

N
PRIME MERIDIAN
English Channel
Berlin
GERMANY
R. Rhine
ALPS
PYRENEES
Madrid
SPAIN
ITALY
Rome
GREECE
Athens

Map 2

N
CANADA
Vancouver
Ottawa
New York
Mississippi River
Mexico City
River Amazon
COLOMBIA
PERU
BRAZIL
ANDES
ARGENTINA
Tierra del Fuego

Map 3

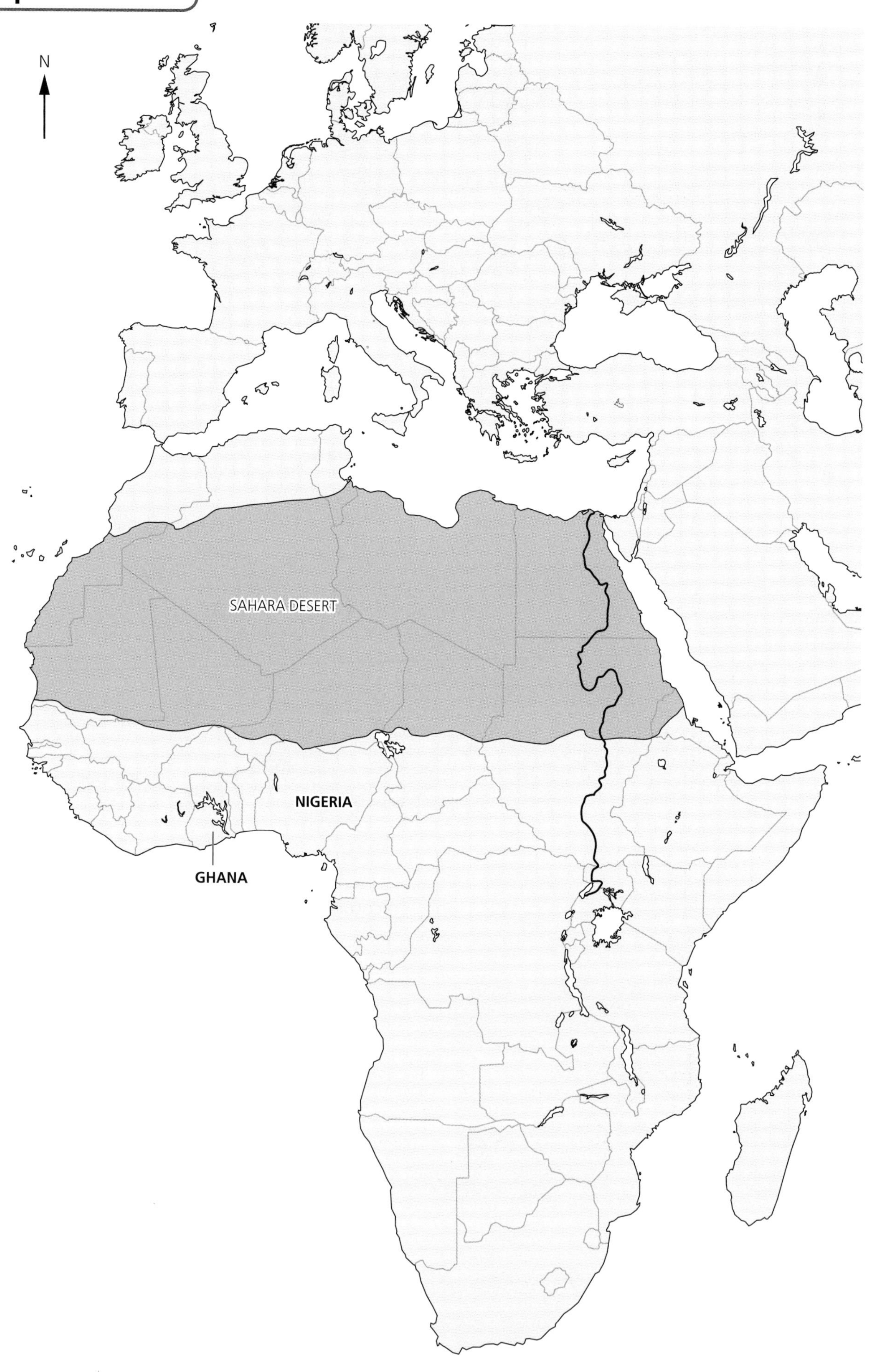

Map 4

Map 5

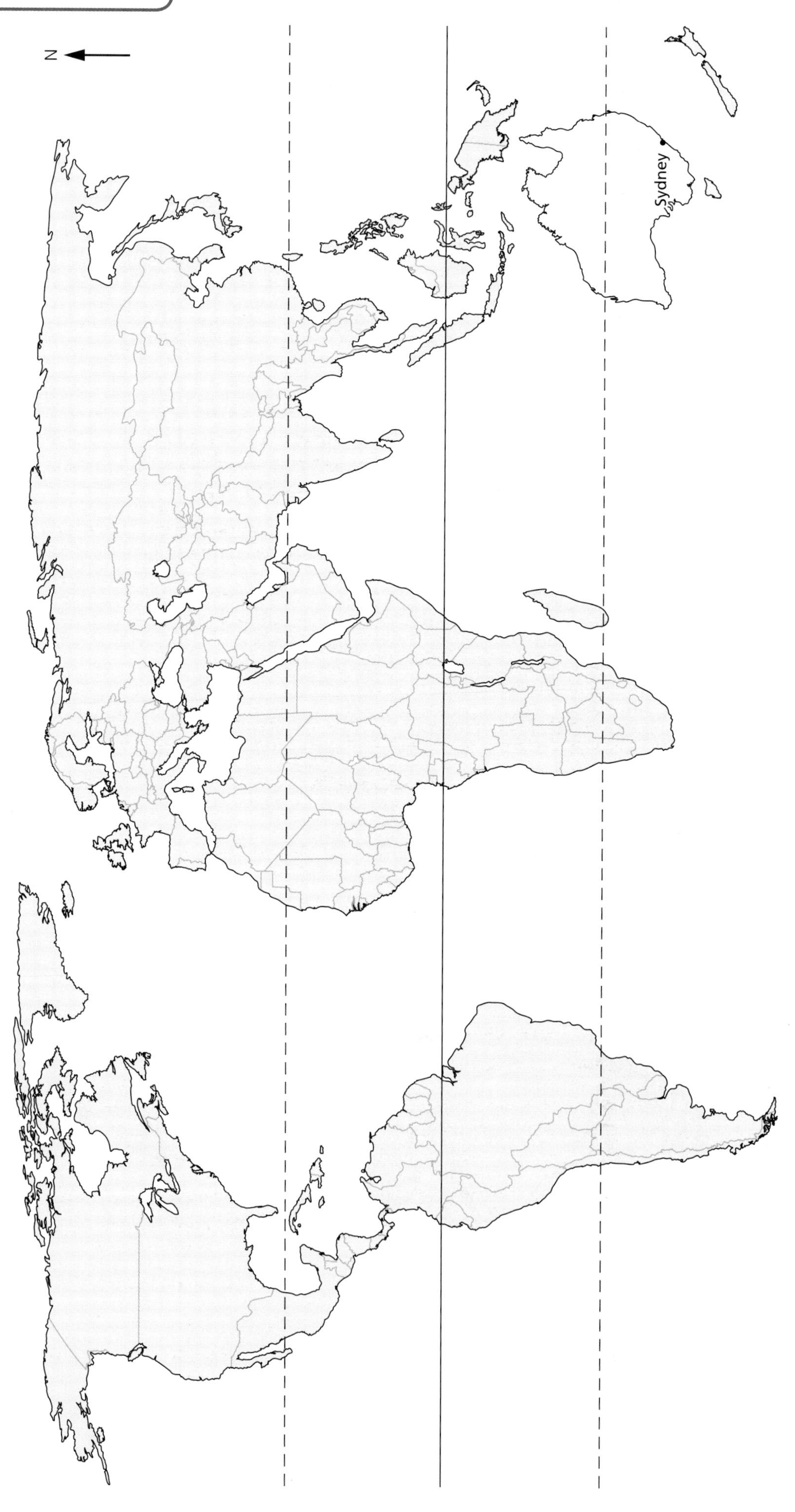

Map 6

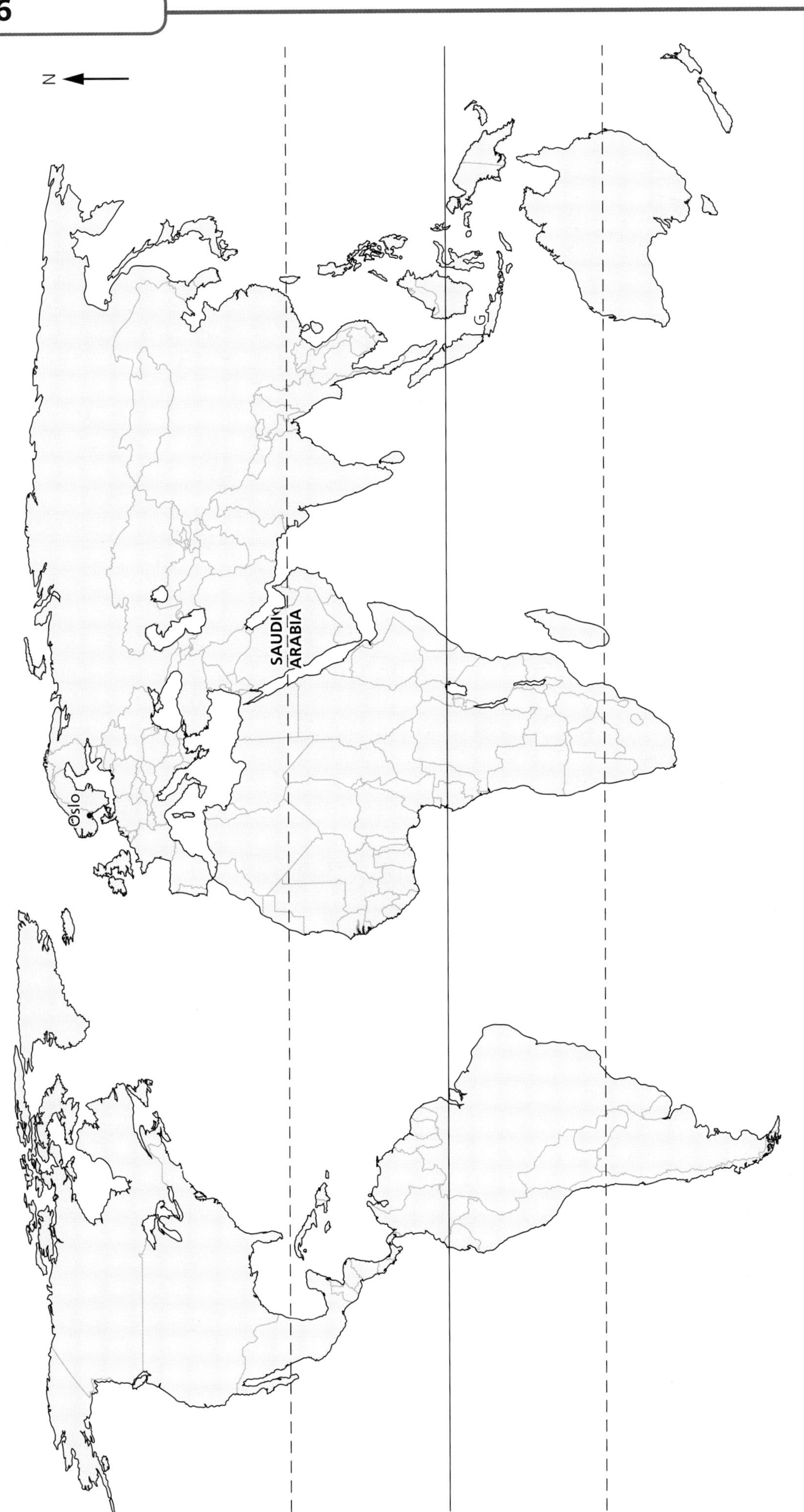

Map 7

N
Arctic Ocean
GRAMPIAN MOUNTAINS
River Clyde
River Severn
English Channel
ALPS
PYRENEES

Map 8